I0481083

Getting into Welding

By:

J. Robert

Getting in Welding

Getting into welding is a great choice of career because some places are always looking for welders. There are different types of welding and different techniques that are available. Some metals are easier to work with than others and some may require extra prep work to avoid contamination. The three main types of welding are M.I.G. (Metal Inert Gas), T.I.G. (Tungsten Inert Gas), and ARC welding. There is another style that doesn't use electricity and uses torch gas called brazing.

The biggest question to ask yourself is do you want to be a welder professionally,just on the side,or a hobbyist. Being a welder can be fun but it can get old quickly to some people. Constant manufacturing day in and day out can get tiresome. Some people don't mind while others do. Some workplaces may have constant challenges and may require you to know how to operate more machines. Others may have you on a line and welding the same thing all day. I'm just bringing this up because of personal experiences and to make you aware of what you're in for before you decide to do welding professionally.

Welding is a great skill to have because where I live and other places across my country have welding positions available. So, if I build up a resume and experience I can get a job anywhere I feel and I can move to any location fitting for me. Also, if you are going to school for something else you can work as a welder until you finish your degree. That way you can make a decent wage instead of flipping burgers and pay for schooling and have some spending money leftover. You can go to school for welding if you wish to because it will help you get a job and better pay up front. You don't necessarily have to go to school to become a welder if you have previous experience or can pass a weld test. Some employers will train new people as well. It all depends upon the business and what standards they require to work at their establishment.

Now, if you're a hobbyist or you just want to pick up another skill it is a good skill to have for fixing things or building things for yourself or others. But make sure you are a good enough welder when doing structural or rotating parts because injury may occur. You can pick up several types of welders for a cheap cost online or at certain stores but remember you usually get what you pay for when it comes to welders. Name brands are always best because cheap brands can generally be disappointing. If you weld very little a cheap brand will be suiting generally but if you're welding weekly then I'd recommend a name brand welder.

Common types of Welding

The easiest welder for beginners is in fact wire feed which is either flux core or M.I.G. This varies for the useage. Flux core wire feed is great for windy outdoor use because it doesn't require a shielding gas to keep oxidation out of the weld. The downfall is that it is dirty to weld and requires a slag hammer to chip the slag off the welding bead when you are finished welding. M.I.G. Welding is the best in my opinion for all around if you want better looking beads. M.I.G. welding does require a shielding gas generally called C-25 which is 25% Carbon Dioxide and 75% Argon. You can run 100% Carbon Dioxide as well but it will have more spatter on the metal. Excess spatter can be caused by bad welding gun angle as well. Although the downfall is that you will have to get a lease or buy depending on your tank size you choose for your welder.

ARC (Stick) welding is an intermediate skill level. There's less control for grip on the ARC welder because you will only have a clamp and a long rod that you make contact with the metal piece. Some do have easy strike arcs but others may bond and act like a magnet to the work piece. Depending on the amperage you may have to shut off your welder or unhook your ground cable to free the rod from the metal joint of your welding. The motion for making the bead is similar to a wire feed welder. ARC welding is similar to flux core because it does use flux to shield the welding rod from oxidation. ARC welders are usually pretty cheap and you can get high amperage welders for a lot cheaper than wire feed welders. Generally 225 amp or bigger is generally the best for a lot of applications. You can either buy an D.C. (Direct Current) or an A.C. (Alternating Current) and D.C. combination welder. If you have the money I'd definitely go for the combination because you can use a wide array of welding rods. A.C. current is great for magnetized steel parts that become magnetized because of friction when hay, feed, water, or even steel on steel start rubbing on the surface. The D.C. current would blow the filler metal out of the weld puddle. Whereas A.C. current alternates and allows it to be welded. D.C. current is also great to have as well. It allows easier starts, less sticking and spatter, welds will look nicer, better for vertical up and overhead style welding. It's easier to learn on DC current. D.C.-reverse polarity offers more penetration compared to A.C. current. D.C.+ is better for welding thin metals. Stick welding is better for contaminated or rusty surfaces compared to other welding types.

T.I.G. Welding is the hardest style of welding. It does require two hands to complete a weld. You can use just one hand if you are fusing metal without adding filler rod but generally you will have to use filler rod to make passes on a

T.I.G. welder. You have to decontaminate a lot more in T.I.G. welding than the other types of welding. Using an acetone type cleaner to clean your work piece. Never weld on rusty metal with a T.I.G. welder it has to be clean. Granted you should always decontaminate your work piece no matter what style of welding you choose. You can clean the surface with a wire brush or grinder with a wire wheel. Sometimes even a sander can be used to clean the surface. T.I.G. welding also requires different shielding gas which is 100% Argon gas. Remember T.I.G. usually runs DC- which is DC reverse polarity which the torch is actually connected to the negative and the ground is positive. This is generally for Mild (Carbon) steel and Stainless steel. Aluminum uses AC current which is usually only available on higher end T.I.G. welders. If you connect the polarity wrong with your T.I.G. welder then you will melt the tungsten very quickly and cause it to ball up.

Brazing is similar to T.I.G. welding but instead of using a shielding gas and electricity it uses Oxygen and Acetylene. Other times people use Propane instead of Acetylene but it doesn't burn as hot as Acetylene which is important for heating metals. Brazing you still have to heat the metal up and then apply the brazing rod to fuse the metal together. Can be used on a lot of different types of metal with different brazing rods. You can braze dissimilar metals together.

Basics on Weld Certification

To pass welding certification tests you need to obviously have enough experience and practice. Then, have a clean work area and make sure the joints are clean where you are welding. Then make a pass on the joint that is no bigger than 3/8 of an inch. Make sure to go at a decent speed. If you go too fast you will get a lack of fusion where there is a pocket of weld that isn't fused into the material in which you are welding. If you go too slow it will cause mounding and an excessive puddle which can cause poor penetration and less strength. After each pass you will have to wire brush or wire wheel the material otherwise it will get contaminated. Obviously this is basic information about this process but there are many online videos and information if you are serious about doing this. Some companies will pay for your weld certification and sometimes welding schools will help you get your certification as well.

Short Arc vs Long Arc

When you are using a M.I.G. welder or a T.I.G. welder you must keep in mind that you have to watch your arc length. If your arc is way too long then you can cause contamination (Also known as pin holes or porosity). Shorter is always better because you always want to shield your bead or puddle that you are making. Remember in welding oxygen is the enemy to your welding puddle or your welding bead. The main thing to look for when welding is you want to be able to look at your welding puddle from the start to the end point and see what you are doing. It's sometimes hard to see it depending on your welding angle but you want to position the nozzle either on a T.I.G. or a M.I.G. system to the point where it is visible to see what you are welding while maintaining a short arc. Granted better welding helmets also have a bigger viewing area in your helmet. So make sure to inspect one before buying a welding helmet. Remember with a short arc you want it extremely short but the shortest possible arc length is always best because shielding gas can do the job properly and your weld will be safe from contamination. The short arc also helps with controlling the puddle and it also creates more heat which is more penetration. Plus, you don't use additional gas because other people have to turn their gas up on their welder because of their bad habit with long arc welding. If you are T.I.G. welding and your short arcing but you end up hitting the tungsten on the metal filler rod or metal and contamination of the tungsten rod occurs. You must use a bench grinder and grind off any metal that bonded to that tungsten. Otherwise you will have an uncontrollable arc. It can also cause a different color light when welding and it can cause contamination to the weld.

Welding wire sizes

There are many types of welding wire sizes such as, .024, .030, .035,.045. They all are used for particular purposes. The .024 is mainly for low voltage welders or auto body work. .030 is used for hobbyists and it's a good general use wire. .035

wire is good for manufacturing and home use that uses a larger amperage welder. .045 is mainly used in manufacturing but in some cases used in home shops but it's not common. Most people usually run .030 or .035 wire. So, now you're asking what is the difference in these wires. Well the thing to keep in mind is that the smaller the wire the smaller the bead and the bigger the wire the bigger the bead. Also, it depends upon your voltage and amperage your welder is capable of doing. If you run too big of a wire it can cause issues with penetration and excessive spatter on your work piece. Too small of wire and if you run too much current can cause spotty feeding in your welding gun. The common wire is flux core which doesn't require shielding gas and solid wire that requires shielding gas. There is another type of wire which is a dual shield. This is a wire that uses flux core and shielding gas. It is used on structural material for construction purposes. It can be used for general purpose as well. It has extreme tensile strength compared to the other types of welding wire. The only downfall is that vertical ups and vertical downs do not have strength when using dual shield wire.

Changing wire spools or fixing wire feed gun issues

Whenever you have an issue and you have to remove the wire out of the welder and back onto the wire spool it's best to cut the wire and take pressure off the feeding wheels. Then, roll the spool so it wraps the wire back onto itself. Make sure to always hang onto the end of the wire otherwise it can create a bird's nest and it can be hepatic and time consuming. Make sure when replacing a wire spool to never let the end go otherwise it will cause the same issue. It's a good idea to remove the tip from the weld gun when feeding your wire.

Wire gun insulators

Wire gun insulators can go bad as well. One case is if you are welding and the nozzle arcs when you touch it against the metal workpiece. What's happening is the insulator is not preventing current from reaching the nozzle and causing it to ground out of the nozzle.

Wire gun tips

Tips are also very important to keep clean and change them when needed. Otherwise feeding issues will occur. Make sure to use the right size tip as well.

Wire gun nozzle

Wire gun nozzles are also very important to keep clean from spatter and slag. If you neglect the cleanliness of the nozzle then it can cause contamination in the welding bead because gas cannot flow through the nozzle properly. You can use a nozzle dip but in certain manufacturing facilities it can cause contamination as well. Using a M.I.G. Pliers to clean out your nozzle is usually the best way but if you have to nozzle dip can work if you are a hobbyist.

Issues with welder feeding

A common issue that causes a birds nest in the feeding wheels in the welder is if the person welding doesn't have their welding lead line straight. If you are welding and your line is coiled up tightly or kinked badly how do you think wire is going to feed through the welding gun.

Shock from welding

Another tip is when welding make sure your gloves aren't wet or soaked. If your gloves are wet it will cause an electrical shock when welding. **Never** take apart a welder without proper knowledge. Some welders have high voltage capacitors that can be fatal if you don't know what you are doing.

Whipping or weaving your bead

A big question is what is the best method for making decent beads when welding. There are a few methods I like. One is the simplest for beginners. If you are welding left to right then whip the welding gun back and forth left to right that way you're welding in the same direction in which you are traveling. A thing to keep in mind is that right handed welders weld right to left and left handed weld left to right. You can also do weaving in the formation of a cursive E. When doing vertical down always take your time and work your way from top to bottom. You can weave it left to right. It's more forgiving if it's heavy metal and your current is set right on your welder then the excessive buildup can fall on the floor. If you

aren't careful you can melt through your metal. Vertical up is better for

penetration but it usually doesn't look great even when experienced welders do

them. Some have turned out looking good but depending on your welder settings

and your material thickness will ultimately determine the beauty of the weld

bead. When welding the biggest thing to watch is your cut and fill rate. Because

when you use a wire feed your cutting into the metal then you are filling it the

rest with the wire being fed from the welding gun. If you go too fast then it can

cause undercut or underfill. Plus, you might have a lack of penetration in the

metal. If you go too slow then it can burn through the metal or cause a cold lap

which is like a big mound of excess bead that looks ugly. Also, if you're not

running enough amperage then it will cause cold laps as well on thicker metals so

make sure when you're buying a welder to identify the metal thickness in which

you can weld. It's all about finding the right balance when welding. Not too slow

but not too fast you're going to want to be constantly monitoring your welding

bead while welding. Which is why it's important to be able to have your arc at a

decent angle so you can see and have proper shielding gas. Remember when you

turn the voltage or amperage up then you need to turn your wire speed up. It all

depends how fast you weld. Some people weld faster and some weld slower. So

adjustments need to be made to compensate for your own speed. Some welders do give you an adjustment table to make adjustments on your welder for the material thickness and wire speed. Although I do suggest to set it at the recommended settings and then adjust if you need to so it best suits your own speed. This does take practice so don't expect to be great your first time. If you get better then weld with your weak hand and try one handed or moving your other hand on a different position on your welding gun. It's always best to become better at different angles because you never know what you are going to be welding on and the location of where you're welding.

Whether to push or pull a puddle

it doesn't make much difference on mild steel. Although I do recommend pushing a puddle because it can create less opportunity for contamination. This is very critical on Stainless and Aluminum steel. You **always** push the puddle or welding bead on stainless and aluminum because you are cleaning and pushing away the oxidation that can form while welding. If you were to pull the bead it would create an opportunity for oxygen to get into the weld puddle and cause it to get contaminated.

Ground a work piece

You can put the ground from the welder onto the table and weld the workpiece but it's still better to have the ground on the workpiece when possible. It makes the welder run more efficiently and avoids interruption while welding. If you do use the table as a ground make sure to have it away from the wall otherwise you can cause the inside of your wall to start on fire. If you ground to the table make sure it is a clean surface and it's bare with no paint because paint is an insulator. Also, a tip when trying to weld on rusty metal or a painted surface if you need to make a tack. The best way to get the wire feed gun to arc if you're having issues is to weld on a scrap surface to get the welder gun hot and then weld on the painted or rusty surface. But only do this on special occasions where it's an emergency or a small stitch or tack. Never do this for structural or important surfaces because the risk of contamination creates weak welds which equals weak tensile strength.

Spot Welding

Spot welding is usually used on auto body because if you run a bead it can warp or burn through the metal panels. It's also good for rusted thin metal. If you ever run into an issue with a burn through or when you keep arcing the metal it keeps disappearing or melting away I may have the solution. The procedure is to run your welder settings on a mid range current or low current. I tend to run the highest current possible so it melts in the metal without globbing. Although, I do have extensive experience in doing this repair so if you're not comfortable you can run a lower current so you don't burn through as easy. Then, all you do is run a spot weld on the edge of the hole. If it burns through then try a different spot. What you want to do is find a spot where it is solid enough where it holds the weld and not burn through. This is a tedious process so patience and small tack or spot welds are important because you are building material to plug the hole. If you do find a solid spot let the spot weld cool so it isn't orange or red in color otherwise it can burn through and you'll make the hole bigger than what it is. Then continue spot welding on the solid spot until the hole is completely filled. Careful not to get brave and run another bead on top of your series of spot welds otherwise you may burn through again depending on the metal thickness. Remember spot welds aren't long; they're about a duration of about 1 second of welding. If you over do it then it will burn through. You can add other types of metal to help fill in but certain applications do require this procedure.

Extraction Welding

Extraction welding is a great way to remove stubborn bolts or broken bolts in certain items. The best way to extract is to get an oversize nut over the piece of bolt and weld onto the nut and bolt so they are as one. Make sure not to damage the hex surface on the nut because you will need a wrench to remove it. Once you weld it make sure to let it cool down. Do not turn the nut with a wrench when the nut is glowing orange or red. It needs to harden so make sure the nut and bolt are cooled down so there are no orange or red heat colors remaining.

You can also use penetrating oil to cool it down. You can also spray it into the threads that are seized in the casting because it helps draw oil into the threads when the metal is heated up although if it's extremely hot like in a red or orange color then it just disperses and doesn't draw into the threads. Once it cools down properly but don't let it cool down too much because heat does help with extraction. You're going to take the wrench and rock the nut back and forth slowly. Inspect the bolt you are removing to see if it's moving at all. This process may need to be repeated because sometimes the nut comes off the bolt. It can take several tries before it comes out. But if you try this and can't get it to come out make sure to stop what you are doing. Your welder may not have enough amperage to be able to get enough penetration.

Extra penetration

Here's what I'd like to address to others on how to get extra penetration when welding on material. If your welder isn't big enough then you should update and get a better welder. But if it's an emergency and you need to get more penetration with your welds on a project then here's a solution. Use an Oxygen-Acetylene torch and heat the area in which you are welding. Don't get it extremely hot. But get it where the metal is a light red color. Never go to an orange color otherwise it can distort and weaken the steel. Then make your pass on the metal. This only applies to Mild steel and Stainless steel because

Aluminum can melt if you try this method. Although there is a shielding gas that can help you in this process with a gas wire feed welder. You need to get a tri-mix which can consist of 66% Argon, 26.5% Helium, 7.5% Co2. Tri-mix can vary in different ratios of these three gases. Helium burns hotter and can make a smaller welder have the ability to have more penetration. This is extremely helpful if your house is wired for 110v instead of 220v. Because the typical max amperage is 140 amps for 110 volts which you will need a 20 amp circuit otherwise you will have to reset your breaker from time to time depending on how much welding you do. So, if you want to get more penetration from your welder and your limited get tri-mix shielding gas. Just remember that your weld setting will be different than the factory specified chart that may be on your welder for setting it up because it doesn't calculate for tri-mix. The only downfall to tri-mix gas is cost but it can be well worth it if you are limited to the size of the welder.

Stitch Welding

Stitch welding is a great way to weld long pieces of steel without warping them. For example if you are welding a 3 inch bead and skipping 4 inches in between and then welding another 3 inch bead. Then alternating from the left side of the work piece to the right because if you continue on the same path then it can overheat and cause the metal to warp. Usually if you're in manufacturing there will be blueprints that show the length of stitch and show you how many stitch welds the material requires.

Rotating welding

Rotating welding consists of pipes that need welding while rotating you are positioning your wire feed gun in a certain location. Your rotator needs to be set at a consistent speed and sometimes they are adjustable. Then your welder needs to be set at the right wire speed. Then you position your wire feed gun at the correct spot and start welding with the rotator. If you are too much in front of the rotator it mounds up and looks bad. If you are too slow then it globs and excess weld splatters when it hits the floor. If you can do it right then it looks continuous and sometimes you won't have to grind your stops and your starts if you are experienced.

Metal Contamination

Metal contamination causes porosity (pinholes) to occur in your welds. Common things such as rust or paint causes the metal to be contaminated. Galvanizing also causes a lot of porosity when welding and creates a stinky work environment, changes the color of your weld, and makes the welding arc unstable at times. The best way is to grind away the surface you're going to weld. Galvanizing is harmful to the body and a welder's tip is to drink Milk to neutralize the galanizing gases.

Along with wearing an appropriate respirator which should be worn when welding.

Gas Contamination

Sometimes shielding gas may be contaminated even from the gas depot. It can cause porosity (Pinholes) in your welds. Make sure your arc is short and properly shielding your weld puddle. Make sure there is no contamination from the steel workpiece like rust or paint and by trying a different piece of clean steel. Make sure your nozzle is clean from any debris or contaminants. If you verify it is indeed the shielding gas then you can exchange your tank but let the gas depot know of the contamination in the tank.

Aluminum Contamination

Aluminum contamination is very common. When welding people often use acetone and wipe the surface and start to weld then in result they get contamination in their weld and they don't understand the reason for it. It's simple, it's usually a two step process. Use a **stainless steel** wire brush exclusively just for aluminum. Don't use a grinder and a wire wheel because it can cause worst contamination because you're forcing aluminum oxide farther into the metal because of heating up the metal and causing the aluminum oxide to bond into the metal. If you have to use a grinder and a wire wheel only use a **stainless**

steel wheel and don't overheat the metal. Grind in the same direction and not back and forth so you can get rid of the aluminum oxide. If you're using the brush in one particular direction away from the weld bead and away from the material because you will have aluminum oxide stuck in the metal. Then use acetone and wipe it all down. Make sure to get rid of the acetone so it doesn't start the work area **fire.** Remember aluminum oxide is similar to the strength of diamonds. So the melting point is around 3600 degrees fahrenheit where aluminum melts about at 1200 degrees fahrenheit. In turn you can't ever get rid of the aluminum oxide contamination during welding because by the time you would get rid of the aluminum oxide you will have a big hole blown through your aluminum material. Where mild or also known as carbon steel you can sometimes get rid of contamination or porosity when you turn the current up but aluminum cannot be done this way because you will melt the material. So, if you see black peppering in your weld beads when you have aluminum then you didn't perform the prepping process correctly.

Health tips with welding

It's always best to have a respirator when welding and welding shops do carry specialty respirators. But they only filter out so much. If you have ever welded aluminum and you are welding near the floor you may know what happens next. Since Argon is more dense than oxygen it will remain near the floor. What happens is the welder will inhale Argon into their lungs and then it will feel like heavy breathing or shortness of breath. The Argon gas will sit in the bottom of your lungs so the best thing to do to remove the gas out of your lungs is to find a chair or a bed and put your head on the floor so your lungs are above your head. Then you're going to inhale and exhale deeply. It takes 15-20 minutes depending on how long you've been exposed to the Argon gas. Trust me it's not a fun time but it is necessary to do even if you have a respirator on it will not stop the gas from entering your lungs. The best way to avoid this is to weld above the floor preferably on a welding table and have an exhaust fan that removes the gas. But make sure the fan isn't positioned too close because it can suck away the shielding gas and cause porosity on the weld bead. **Always** wear protective eye, ear, and lung protection equipment to protect your health.

Choosing a Welding Helmet

Now there are many types of welding helmets but there is a main style I suggest. Auto darkening for one is a great helmet for beginners or professionals. You can use a straight dark welding shield but auto darkening is great for seeing your starts and stops and see where you are starting on your weld. Granted many professionals used straight dark welding helmets but I still prefer auto darkening. Now depending upon how much visibility you want generally the low end auto darkening is going to have a small viewing lense. Usually the more money you spend you will see an increase in the viewing lense as well. So make sure you do your research and identify the size of viewing lense you want. Remember the bigger the lense the easier it is to see and less likely for strain and neck aches from trying to see what you are welding on. Depending on the application such as under vehicle welding and hard to reach areas. If you are welding on a table or a comfortable surface then it's not as big of an issue. Not that it's required but it makes it easier on yourself. If you can't afford it that's fine but I do recommend getting a bigger lense later if you continue welding. Generally you will see features on your welding helmet such as sensitivity, shade adjustment, and delay. Some even have a grind mode which allows the helmet not to darken when grinding and allow it to be a protective shield for your face. Sensitivity is to adjust how sensitive the helmet is to light the higher the sensitivity the more it's prone to auto darken. So make sure your sensitivity isn't too high otherwise it will auto darken when you don't want it to. When it comes to shade adjustment it really is recommended on your amperage. The larger the number shade the darker it is. The old school welding helmets were a straight shade 10 but it really depends upon your amperage that you are welding. The more amperage the brighter the U.V. light. Remember looking at welding is like looking at the sun which will cause flash burn so it's important to get the right shade. Now the A.W.S. (American Weld Society) is saying that shade 11 is actually recommended instead of shade

10. Shade 11 is good for 60-160 amps, shade 12 for 160-250 amps, Shade 13 for 250-325 amps, shade 14 for 325-500 amps. Now most welding helmets will have a shade 13 for the maximum size shade but general welding for most will not require shade 14. Delay control allows the length of the welding helmet to be darkened after you weld. Usually a half a second to two seconds of delay on most auto darkening helmets. So, if you tack welding then a short delay is necessary but if you are welding for longer periods of time a longer delay is better for your eyes. Adjust it for your own preference.

Welding cast iron

Ferro-Nickel Rods for stick welders are the best for welding on broken engine block castings and exhaust manifolds. The Nickel flexes with the cast iron to prevent the weld from cracking.

Welding outdoors

When you weld outdoors you can use a gas shielded welder but make sure there is no wind present otherwise you will get porosity (pinholes) in your welds. Also, never have a fan blowing on you regardless how hot the weather is because it will cause porosity as well.

Best way to get rid of Porosity

Porosity usually is a pain. Whether you forgot to turn the welding gas back on, you weld on contaminated metal, or a breeze or fan hits your area while welding. Now, if this occurs you will most likely have to grind those pin holes out of the welding bead. Sometimes you can run another bead and get rid of the issue but most times if you would run another bead across the top of the porosity bead it will make it extremely large or more pin holes will occur and then you have twice as much to grind. If you want it done the best way then grinding is my preferred method.

T.I.G. Welding Tips

T.I.G. is the most difficult type of welding for most people. The most important thing is to keep a short arc form. The distance your T.I.G. electrodes should be equivalent to the thickness of your steel. For i.e. if your welding 1/8 inch steel then your tungsten electrode should only be 1/8 inch from the weld joint. Another thing is your torch angle. If your torch angle is way too much then it will cause issues when welding. Also if your puddle is a C shape when welding then your arc may be too long and your torch angle is too much. You want it to look like a teardrop shape when welding. Make sure you are also shielding your filler rod as well. If you have too much arc length it can cause the weld bead to dramatically increase in size as well. The filler rod should flow like water into the weld joint not blobbing. Remember if you don't have a puddle formed then you aren't welding. Depending on how many amps you are running, usually 100 amps is a good base on most situations. But if you only have an 80 amp T.I.G. welder it may take longer for a puddle to form. Simply heat the metal and go back and forth on the joint until the puddle touches both points of the joint. Sometimes you may get away with a smaller size T.I.G. welder but don't over push your applications. Make sure to get the appropriate size welder for the thickness in which you plan to weld on. If your right handed you will weld right to left and if your left handed you will weld left to right using the same principle as other welding types. When starting your arc with a live strike T.I.G. welder simply strike your arc similar to a stick welder but requires little movement. If your torch sticks then maybe your tungsten isn't sharp enough or your welder type settings are off if you have a T.I.G. and Stick combo welder. If you have a peddle style tig welder you want to slowly hit the foot pedal very slowly. It can be sensitive so it's basically like a granny driving a car. If you apply too much pedal then it can cause too much heat and blob the material. Some pedals are adjustable in amperage as well. Get the feel for the pedal and practice on some scrap pieces to master it. When you get your arc started on a T.I.G. torch your going to tip the torch angle so it's about 15-25 degrees. You want to be able to see what you are welding. A clear gas lens always helps if you're beginning as well. You're going face the tungsten electrode in the direction of travel. Basically you are going to feed the puddle with a filler rod. Always bring the filler rod at the lowest angle possible.

Don't let the filler rod hit the tungsten otherwise it will contaminate it and you

will have to grind the tungsten rod again. Remember to look up online the

tungsten degree for your application. Some may require 16 degrees which is

extremely sharp, 32 degrees, 60 degrees, or even blunt which is commonly used

for aluminum. Remember Thoriated electrodes do have little radioactive

material. It can be possibly hazardous when grinding so wear a respirator. When

welding it isn't proven to have any health effect on the person welding. Welders

have been using it for many years without health hazards as well. The benefits

are that the Thoriated tip does last longer than most T.I.G. electrodes. But if you

are wanting an alternative I do recommend 2% Lanthanated for all types of

metals. There are many other good types of electrodes for T.I.G. welders but

make sure to do some research to find which is best for your application. A

sharper angle on the tungsten grind also helps with more penetration. For i.e. a

16 degree tungsten electrode will have more penetration than a 32 or 60 degree

electrode. Another helpful tip that applies for stainless and mild (carbon) steel is

that a sharp angle like a 32 degree or 16 degree electrode helps with an easier

strike with a live strike arc welder. Make sure your tungsten electrode isn't

sticking out too far. Dependinging on your gas lense size. If you have too much

electrode hanging beyond the cup then it can cause porosity because there is no gas shielding the weld puddle. But in certain cases you may have to have more electrodes sticking out because the joints are in hard to reach areas. Oftentimes being tight fitting pipe joints. That is why gas lenses are so important to get the right one for your application. So how much stick out is acceptable? To answer that question a number 5 gas lense has 5/16 inch diameter hole which allows about a 5/16 inch stick out from the bottom of your gas lense. A number 8 gas lense is a 1/2 inch diameter hole and allows 1/2 inch stick out. Seeing the pattern? Common gas lense sizes range from numbers 4-16. Which a number 4 is a 1/4 inch and a number 16 is a 1 inch. Certain specialty types of T.I.G. torches go to a number 24 which is 1 and 1/2 inch.

Conclusion

Things to remember when welding:

Short Arc is always a must for the best welds otherwise porosity will occur and in T.I.G. welding your puddle will not flow smoothly. Also make sure not to have too small of an arc when T.I.G. welding otherwise it can contaminate the tungsten.

Always clean your metal from contamination because contamination will cause porosity and weak welds.

Slow and steady. Remember welding too fast isn't good for anything because of lack of penetration and filling the welding joint. Slow down when welding and when you're ready then you can speed up a bit but it's better to be slower than too fast while welding.

Always consider safety for your lungs, eyes, and ears when metal working. Always be safe!

Make sure to research the electrodes or wire diameter that best suits your application.

Be considerate when welding. Don't flash others around you if you don't have to. Use your gloves to shield the arc light or get a divider that is approved for welding.

Remember welding produces U.V. rays make sure to wear approved protective clothing and gloves to protect your skin from active burns and U.V. Ray burns.

Practice is the only way you will ever get better. Practice is the only way you are going to get better. If you get good with welding on your dominant hand try your weak hand and if welding stick or M.I.G. try one handed. Always improve yourself so you can handle hard angles and improve your skill.

When there are two different metal thicknesses weld on the heavier thickness first.

Never use 100% Argon gas when M.I.G. welding mild (carbon) steel or stainless steel. Only use 100% Argon for M.I.G. welding Aluminum otherwise your arc will be unstable and weld beads won't turn out decent.

Make sure if you have someone to guide you that they themselves are experienced welders they can help you. Don't take advice from those who have no knowledge of welding.

"A blind man is one who can't see the truth but the man who has seen the light carries on with the truth."

John 9:39

www.ingramcontent.com/pod-product-compliance
Lightning Source LLC
Chambersburg PA
CBHW080630220526
45467CB00011B/3444